YOUR KNOWLEDGE HAS VALUE

AF144309

- We will publish your bachelor's and
  master's thesis, essays and papers

- Your own eBook and book -
  sold worldwide in all relevant shops

- Earn money with each sale

Upload your text at www.GRIN.com
and publish for free

Sebastian Kerski

Artificial Food Colors as Enhancers of Hyperactivity in Children

GRIN Publishing

Bibliographic information published by the German National Library:

The German National Library lists this publication in the National Bibliography;
detailed bibliographic data are available on the Internet at http://dnb.dnb.de .

Imprint:

Copyright © 2014 GRIN Verlag GmbH
Print and binding: Books on Demand GmbH, Norderstedt Germany
ISBN: 978-3-656-86587-2

This book at GRIN:

http://www.grin.com/en/e-book/286415/artificial-food-colors-as-enhancers-of-
hyperactivity-in-children

GRIN - Your knowledge has value

Since its foundation in 1998, GRIN has specialized in publishing academic texts by students, college teachers and other academics as e-book and printed book. The website www.grin.com is an ideal platform for presenting term papers, final papers, scientific essays, dissertations and specialist books.

Visit us on the internet:

http://www.grin.com/

http://www.facebook.com/grincom

http://www.twitter.com/grin_com

Artificial Food Colors as Enhancers of Hyperactivity in Children

Authored by: Sebastian Kerski

ZUYD University of Applied Science

Submission date: November 03, 2014

Attention-Deficit/Hyperactivity Disorder, short ADHD: Who has not yet heard this term, especially when it comes to children? Over the last decade the number of children diagnosed with this disease has risen significantly (in the United States from 7.8% in 2003 to 11.0% in 2011. Centers for Disease Control and Prevention, 2014). But although this diagnosis becomes more and more common, there is still a lot that we do not know about its causes and right treatment. Not only does the right medication help to guarantee children a normal development, but research also found out that the right diet can have a positive effect on the symptoms of hyperactivity. Artificial food colors in particular have been named again and again in this context.

This essay focuses on the most important research concerning the field of artificial food colors and their impact on children with hyperactivity: There are several studies which suggest a link between an increase of hyperactive behavior and certain artificial food colors, and while some countries have already taken steps to remove the tested colors from child nutrition, others still deny their potential danger. This shows how important a more thorough investigation of this topic is and that the restrictions concerning artificial colors in food for children should be revised everywhere.

One of the first researches on the effects of artificial food additives on hyperactive children was Dr. Benjamin Feingold's study in the early 1970ies. He tested hyperactive children by eliminating certain foods and artificial additives from their diet and stated that these children showed less symptoms of ADHD (Kanarek, 2011, p. 387). Feingold's method became quite popular and many parents of hyperactive children decided to trust in his "elimination diet" (Kanarek, 2011, p. 387). Even today the Feingold association exists and recommends a diet free from certain foods and artificial additives for children diagnosed with ADHD. You can join the association on www.feingold.org.

Nevertheless, Feingold's method soon met many critics, whose studies refuted his rigid program. These studies found out that certain foods and additives can have a negative effect on ADHD patients but an elimination diet does not necessarily improve the patient's condition. Only some of their probands had less symptoms of ADHD when they did not consume the tested artificial food additives (Williamson, 2008, p. 4).

Consequently, one can say that Feingold was a pioneer in the research on artificial colors and food additives and their influence on hyperactive children, but his suggestions for a treatment by eliminating certain foods from their diet had to be revised.

This task was undertaken by many researchers until, in 2007, the Southampton study provided results that were significant enough to cause the British government to ban artificial food colors from the British market (Kanarek, 2011, p. 389). Unlike the Feingold study, this one concentrated on artificial food additives. Two drinks were given to children who were not diagnosed with ADHD. The first mix contained the artificial dyes sunset yellow, tartrazine, carmoisine and ponceau 4R, while the second one contained sunset yellow, quinolone yellow, carmoisine and allura red. Prof. Jim Stevenson from Southampton University summarizes the results as follows: "This has been a major study investigating an important area of research. The results suggest that consumption of certain mixtures of artificial food colors (. . .) are associated with increases in hyperactive behaviour in children" (Food Standards Agency, 2007).

But – as well as the critics of the Feingold study – this research also clarifies that hyperactivity cannot be cured or necessarily be lessened by eliminating the tested substances from children's diets. Still, the results are alarming enough to indicate that artificial colors can be quite dangerous.

Nonetheless, not all institutions admit that artificial colors are a problem. The U.S. Food and Drug Administration (FDA), for example, denies a negative effect of these ingredients and explains their importance in the manufacturing of food, because they ensure a certain standard in its appearance. Additionally, the FDA states that only certified colors are used in the food industry and that these have to fulfill a very strict standard. Therefore, they say that food with artificial color additives can be consumed without worrying about negative effects (U.S. Food and Drug Administration, 2010).

Besides, the FDA even comments on the findings of researches on food additives and their effects on children diagnosed with ADHD: According to the FDA homepage the studies undertaken until 2007 do not "substantiate a link between the color additives that were tested and behavioral effects" (U.S. Food and Drug Administration, 2010).

The statements on the FDA homepage, though, contain certain inconsistencies. First of all, it is interesting that the homepage was last reviewed in 2010, after the Southampton study was published, but it does only mention a research that denies a link between artificial food dyes and hyperactivity in children. Furthermore, it states that the European Food Safety Authority (EFSA) has the same opinion as the FDA. But this is not true, as the homepage of the EFSA contains a statement on the Southampton study. It says that, as a consequence of the findings of this research, the EFSA changed the acceptable daily intake (ADI) for the artificial dyes quinoline yellow, sunset yellow and ponceau 4R. This shows that recent research has not only convinced the British Food Safety Authority, who, in 2009, reacted to the Southampton findings by banning artificial dyes from food made for children, but also alarmed the European government. Hence, one can say that the data used by the FDA is not quite up to date. Some might suggest that this "lack of information" might be intentional, as it is quite convenient for the argumentation of the whole homepage, which states that there are no negative effects of artificial food additives and dyes.

Also, we have to ask ourselves, if the actions taken by the EFSA are actually enough or if the British government did the right thing. Even if the FDA tries to convince us that artificial colors are vital for the food industry because they are, for example, needed in "'fun' foods" (U.S. Food and Drug Administration, 2010), we could probably also live with mint ice cream that is not green. It would still taste the same.[1] If we are ready to get over the loss of some color in our food, we could make eating much safer not only for us but especially for those children who suffer from all the symptoms that come along with the attention-deficit/hyperactivity disorder.

In summary, we have seen that from the early 1970ies on research on the effects of artificial food colors on children with ADHD has changed and developed. Feingold, who proclaimed a consequent elimination diet, was refuted to a great extent. Later research was more specialized on artificial dyes and even evoked a change in politics concerning the food industry. But still, there are countries like the USA, whose controlling institutions deny a link between artificial colors and hyperactivity. The problem may be the fact that research, so far, has not been able to clearly identify the proceedings of these additives in children's bodies and especially in children suffering from ADHD. In this respect Williamson (2008) is right, when he claims that we need more research on this topic (p. 6). If a further study could find

---

[1]   The FDA homepage states that we actually need artificial color additives for example „to provide color to colorless and ‚fun' foods. Without color additives, colas wouldn't be brown, margarine wouldn't be yellow and mint ice cream wouldn't be green. Color additives are now recognized as an important part of practically all processed foods we eat." (U.S. Food and Drug Administration, 2010)

out what food dyes actually do to the human brain, even the FDA would have to admit that measures have to be taken.

References

Kanarek, R. B. (2011). Artificial food dyes and attention deficit hyperactivity

disorder. *Nutrition Reviews, 69*(7), 385-391.

Williamson, C. S. (2008). Food additives and hyperactivity in children. *Nutrition Bulletin, 33*,

4-7.

Centers for Disease Control and Prevention (2014, September 29). *Attention-

Deficit/Hyperactivity Disorder (ADHD)*. Retrieved 2014, November 03, from

http://www.cdc.gov/ncbddd/adhd/data.html

European Food Safety Authority.(2014, July 15). *FAQ on colours in Food and Feed*.

Retrieved from http://www.efsa.europa.eu/en/faqs/faqfoodcolours.htm 4-7.

Feingold® Association (2014, August 11). *Benefits of Joining the Association & Buying the*

Retrieved September 07, 2014, from http://www.feingold.org/benefits.php

Food Standards Agency. (2007, September 11). *Agency revises advice on certain artificial*

colours.Retrieved from http://webarchive.nationalarchives.gov.uk/20120206100416/

http://food.gov.uk/news/newsarchive/2007/sep/foodcolours

U.S. Food and Drug Administration (2004, November; revised 2010, April). *Overview of*

Food Ingredients, Additives & Colors. Retrieved September 07, 2014, from

http://www.fda.gov/food/ingredientspackaginglabeling/foodadditivesingredients/ucm0

94211.htm